THE ZAMBIT LINE DREAM

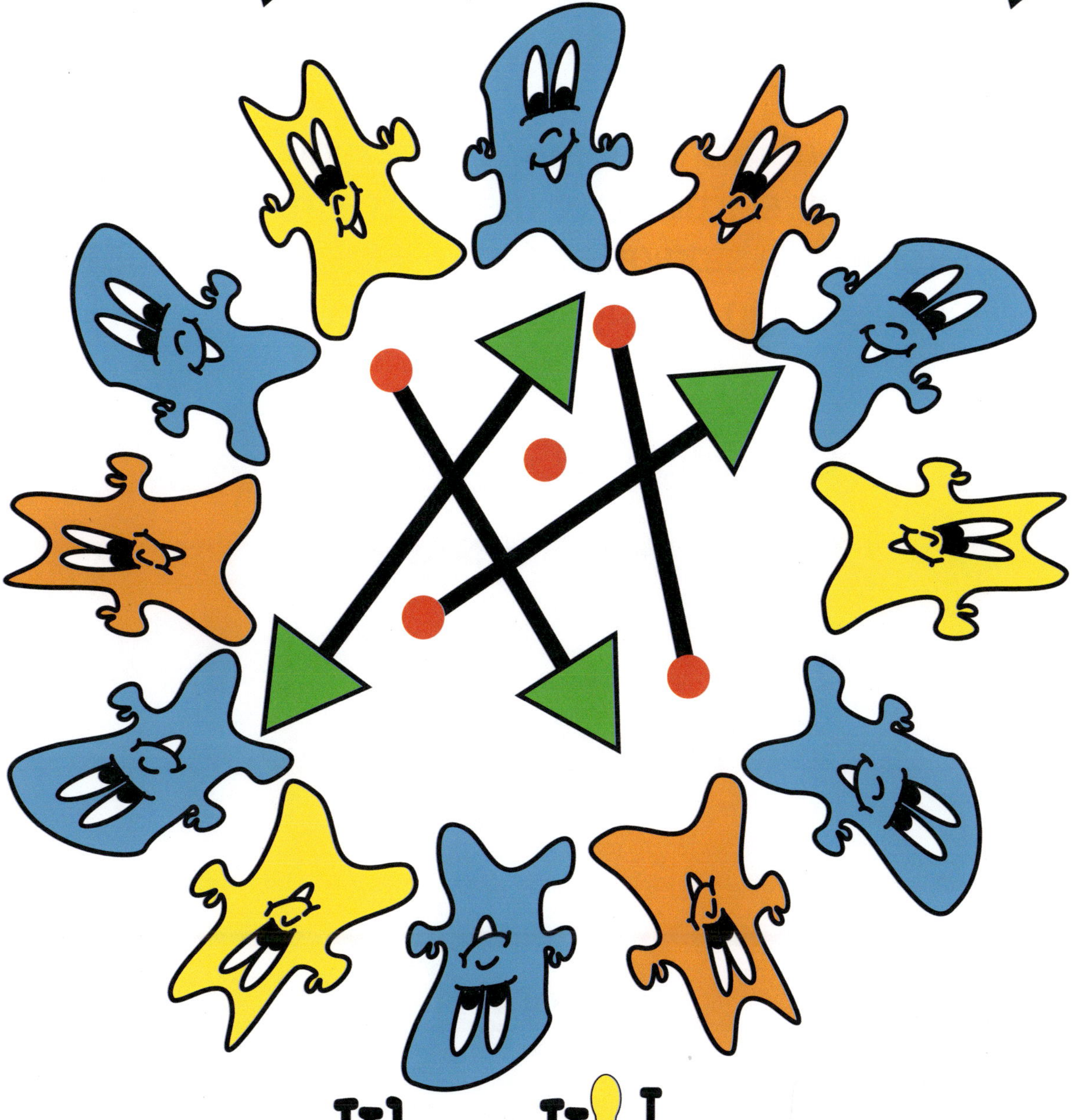

KlevaKids

AF812266

Published by KlevaKids.com Inc

New York, New York

For more titles,visit our website at www.klevakids.com

KlevaKids

KlevaKids

Presents

THE ZAMBIT LINE DREAM

This story introduces the basic concepts of lines in an enjoyable format.

The above statements are purely informational. The purpose of the book is not to teach, but to expose concepts. Read the book to a child and just allow the story to be enjoyed.

It was a pile of stuff,

a real big mess,

that was causing

a whole lot of stress.

The math symbol contraption arrived

quickly on the scene.

It made that pile of stuff

orderly and clean.

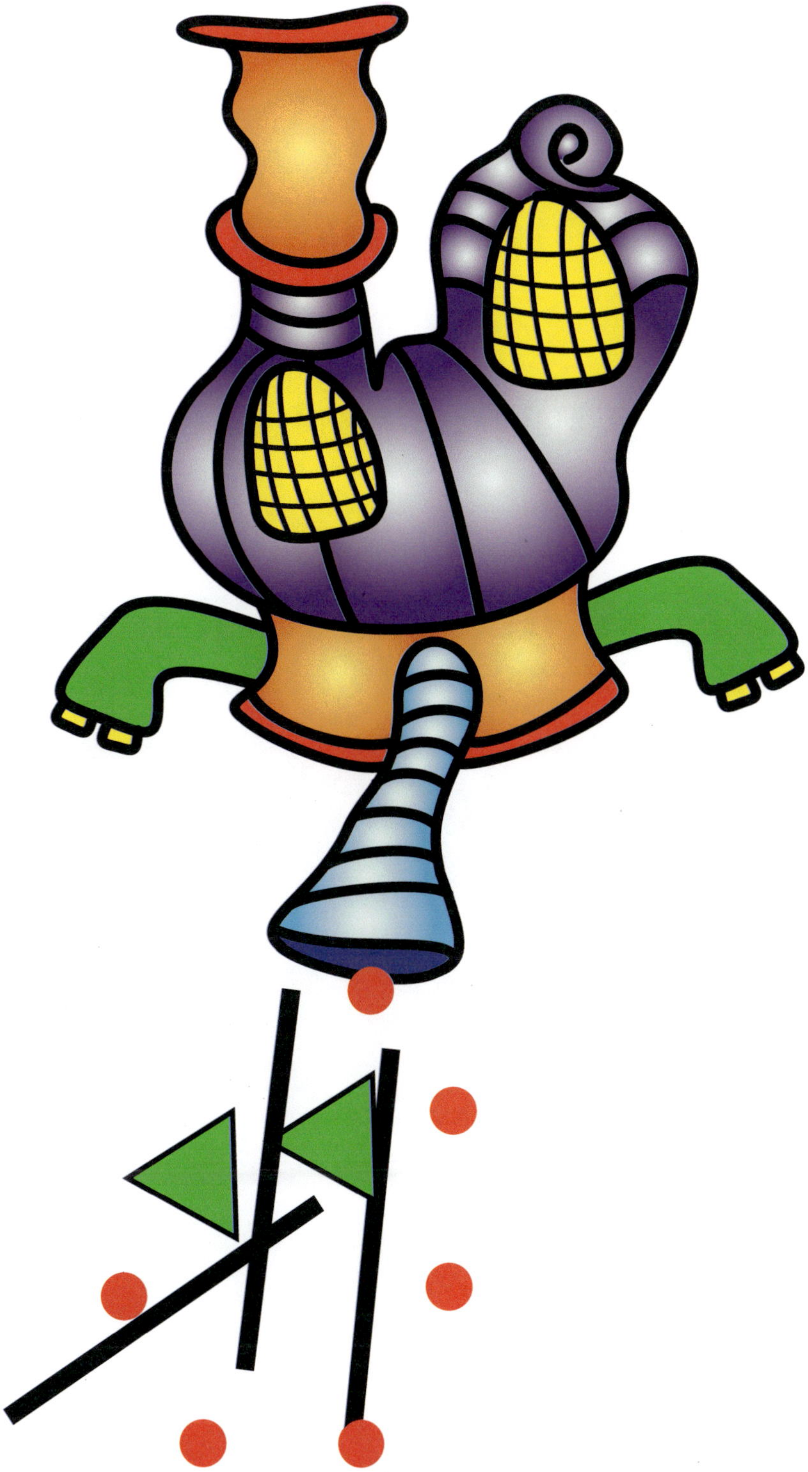

The contraption hissed and puffed.

Something came out like a shot.

The first symbol to appear

looked like a dot.

Next came an arrow

neatly pointing right.

It rose from the stack

and gained a bit of height.

Emerging from the stack

was a left arrow.

It had one end

that was very narrow.

A line segment came out.

It was straight and true.

I had seen one before.

It was something that I knew.

The contraption was leaving,

that was very clear.

It would be back again,

so I had no fear.

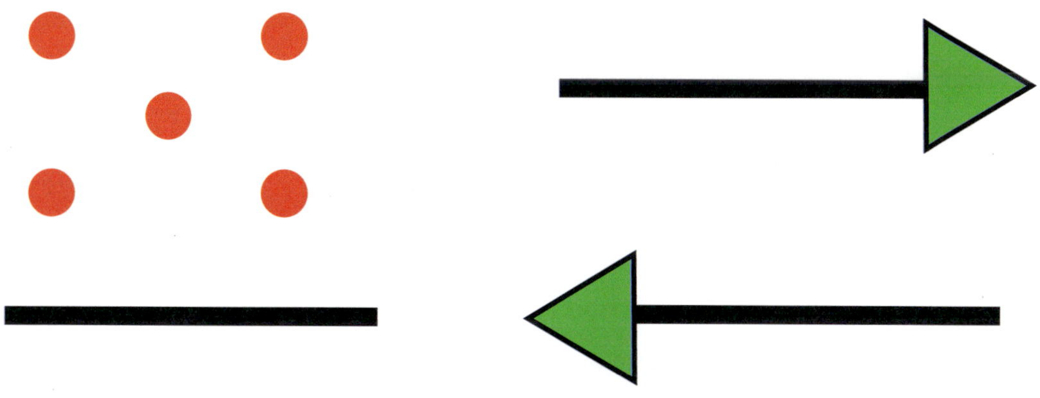

Suddenly it flew closer,

at such a fast clip.

I was happy to see

that it was the geometry ship.

Math Zambits came out

and did a real neat thing.

They said what the symbols represented

after forming a ring.

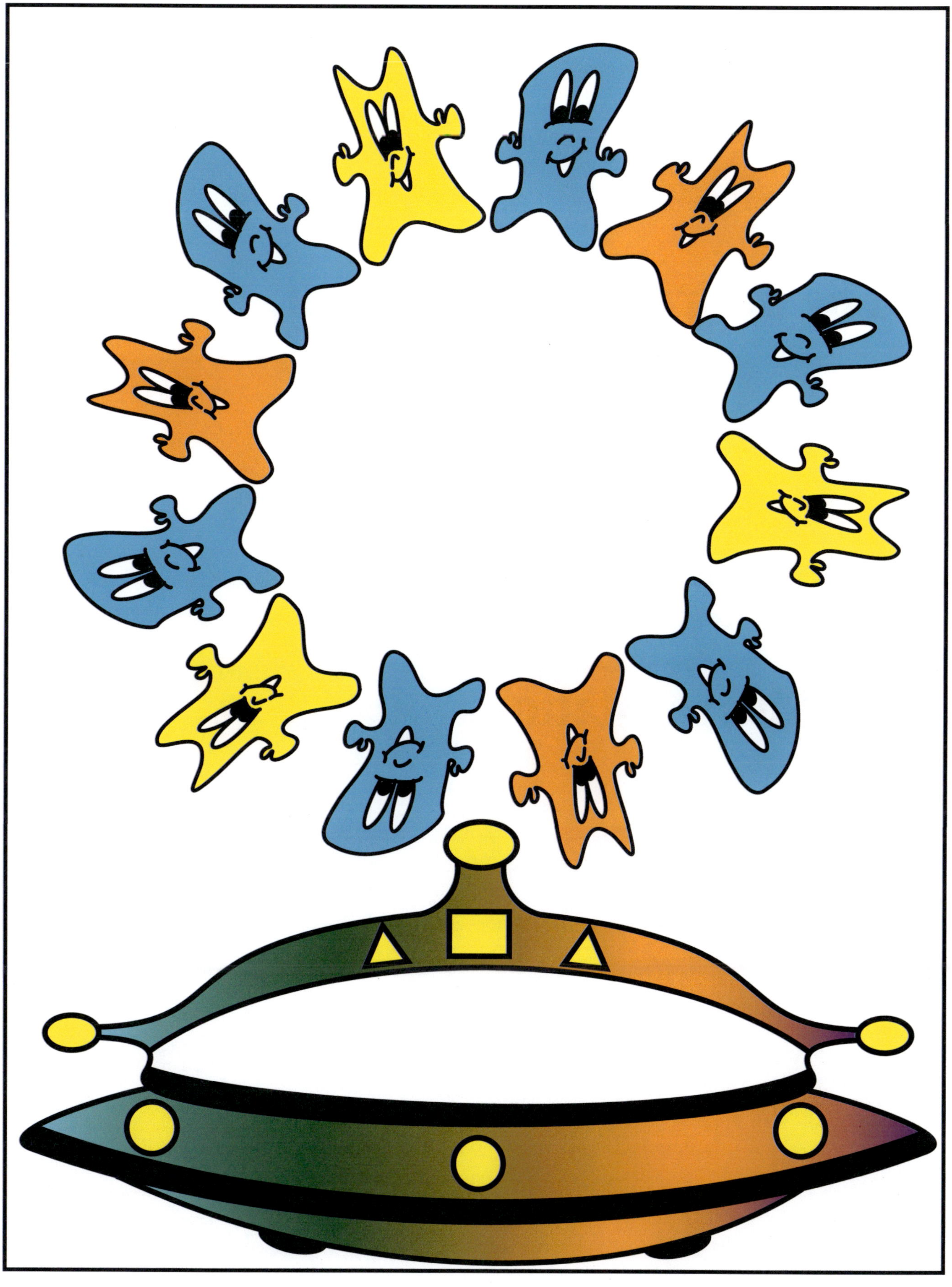

The dot represents a point,

that is a specific location.

It has no height or width

in this translation.

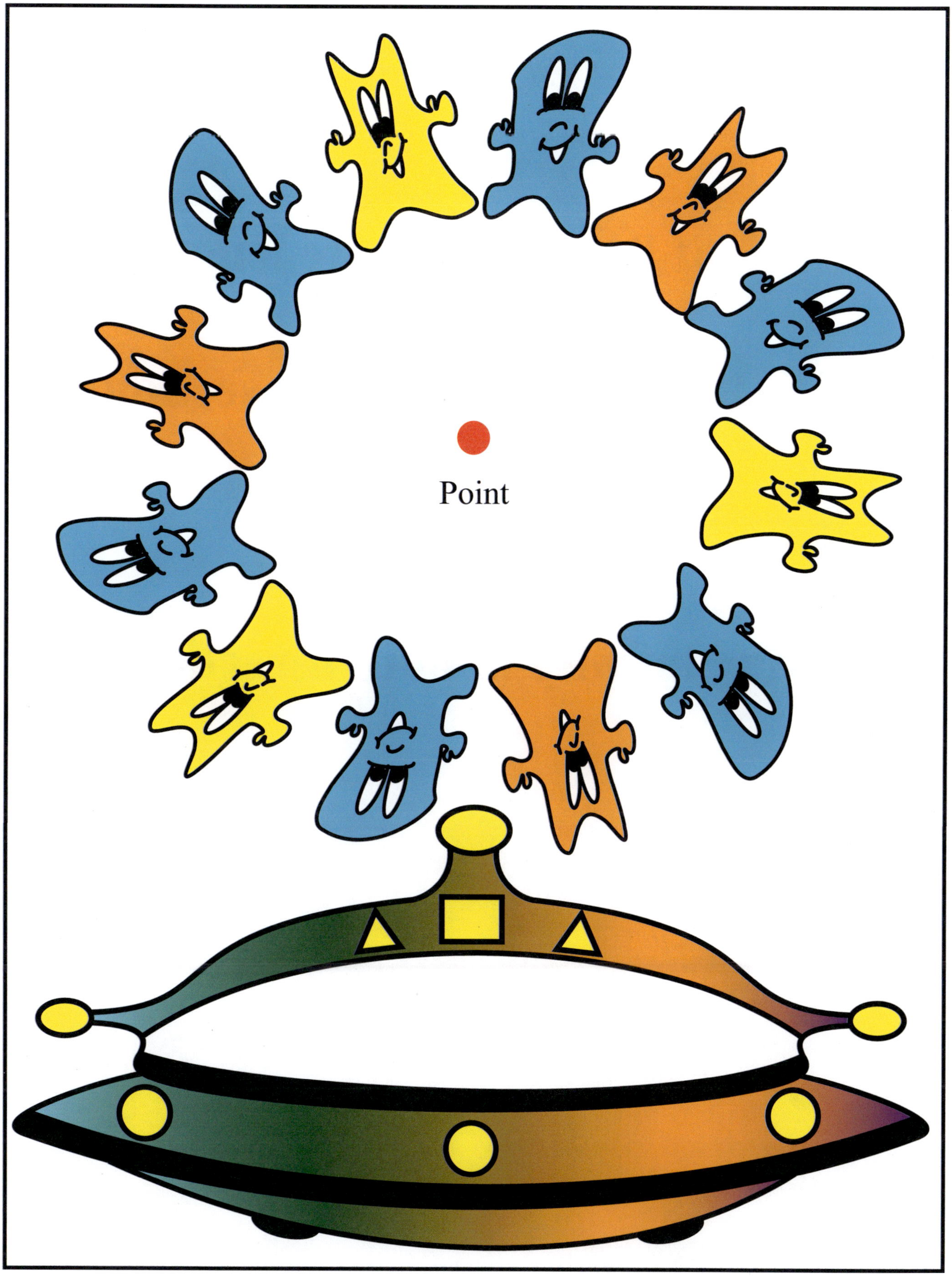
Point

A straight line has infinite length and neither width nor even sections. It is represented by arrows showing infinity in both directions.

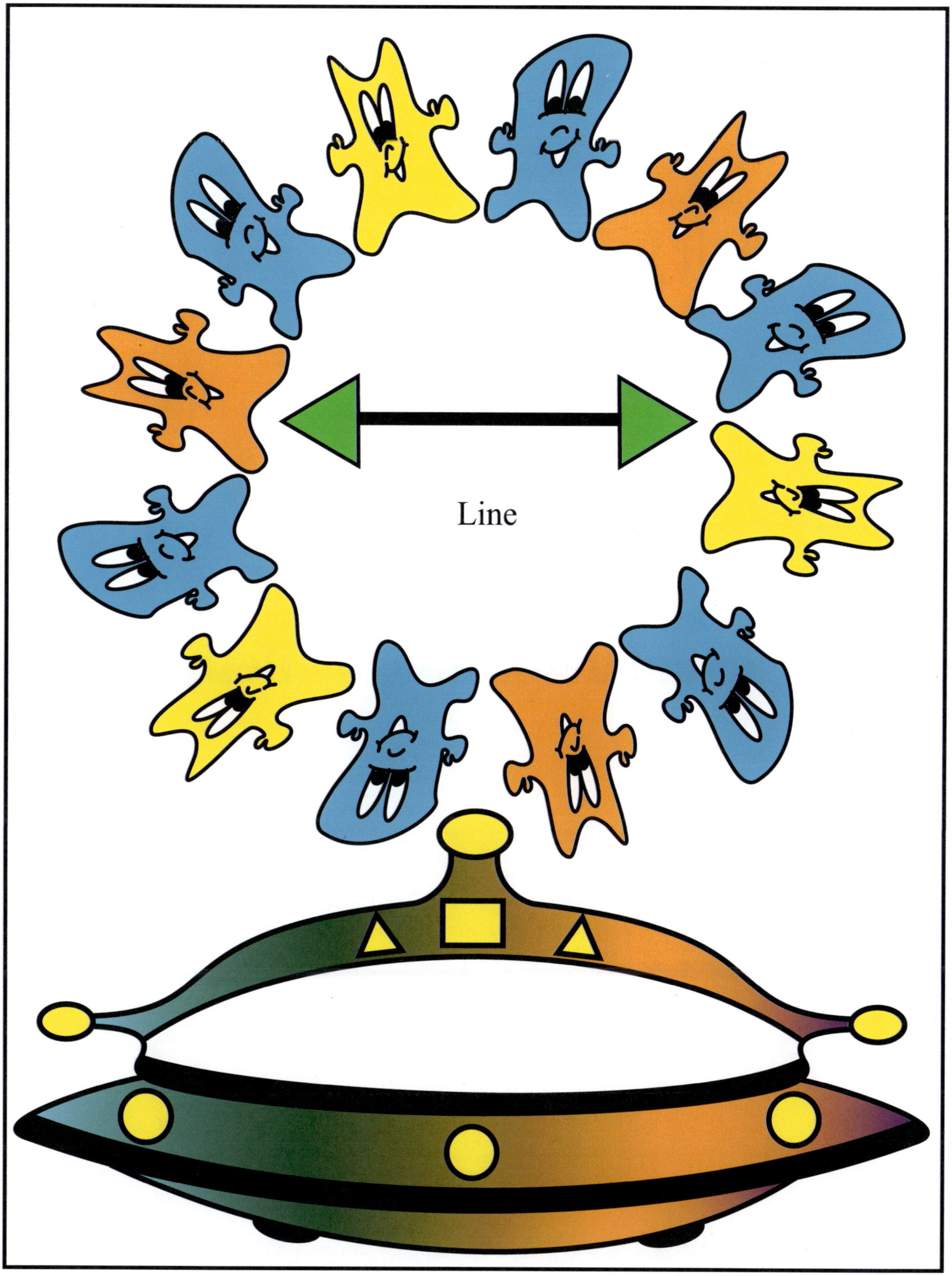
Line

A straight line segment

is often called a line.

The line portion between two points

is the actual definition, by design.

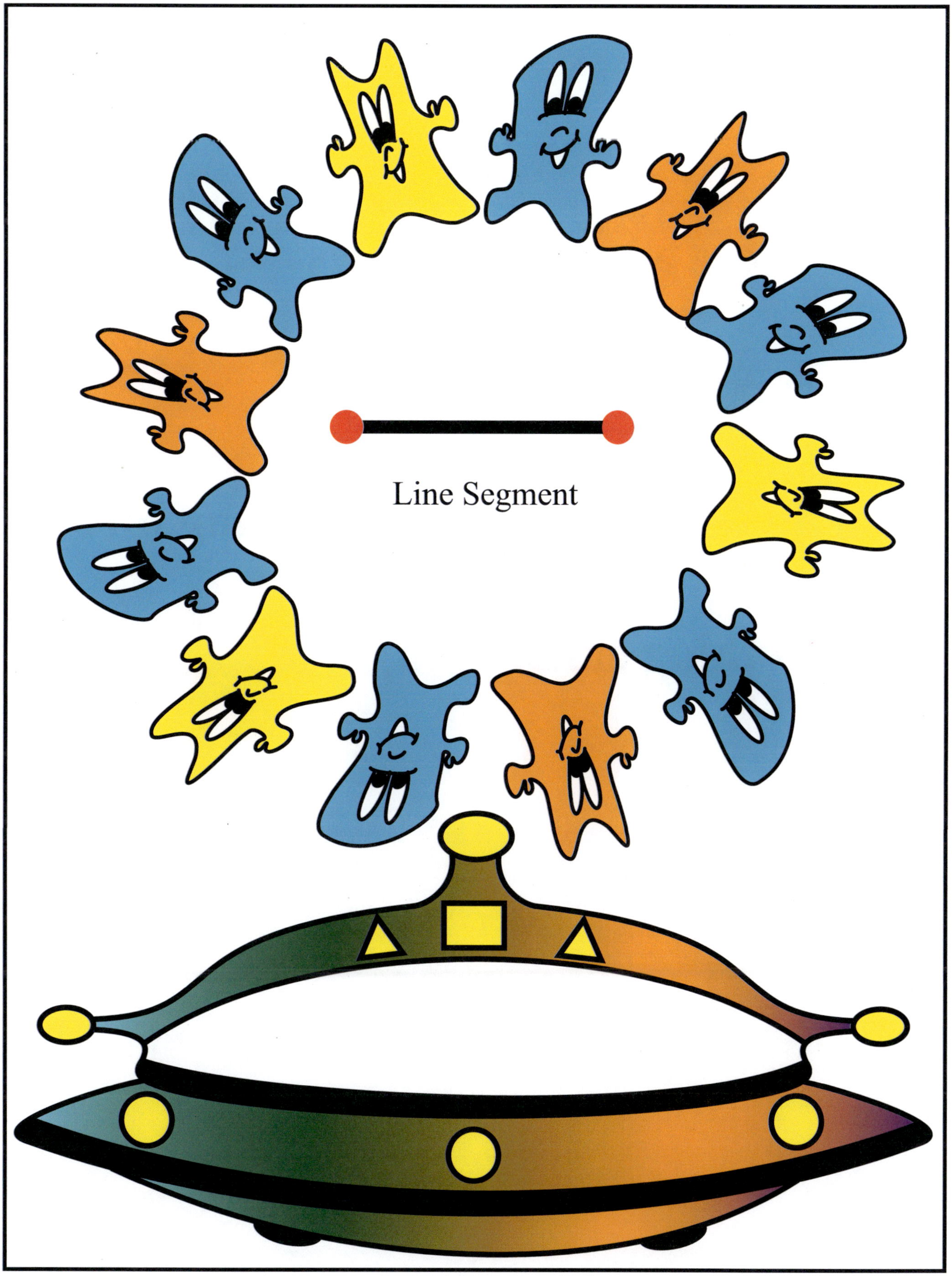
Line Segment

Now there are these arrows

that can point either way.

Starting at a point and extending forever

is something called a ray.

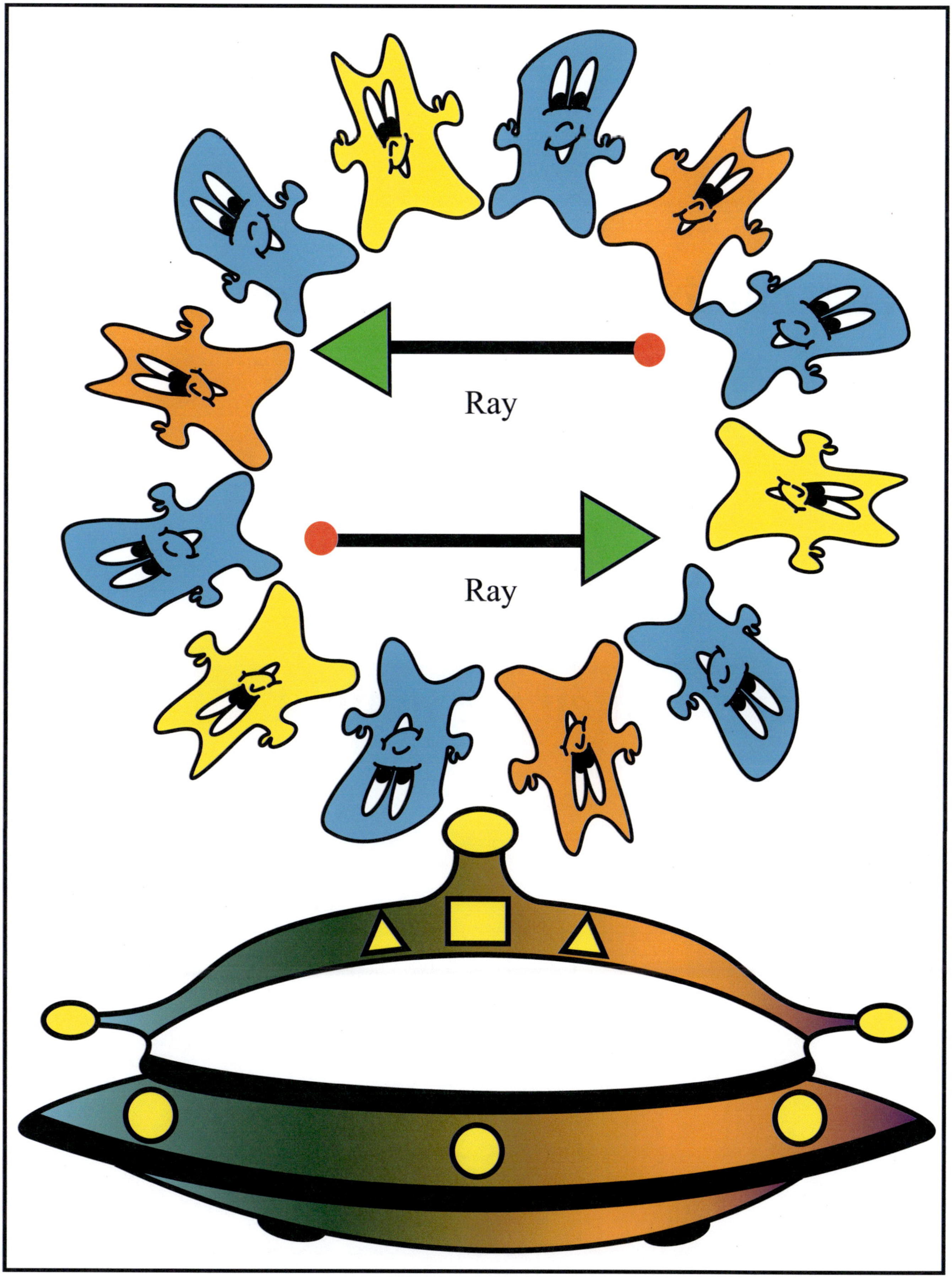
Ray
Ray

The Zambits got in their ship

and took off on a flight.

I was just so happy

that they visited me tonight.

I love to dream of Zambits.

That's exactly what I said.

I like to keep this line stuff

right inside my head.

THE END

Point

Line

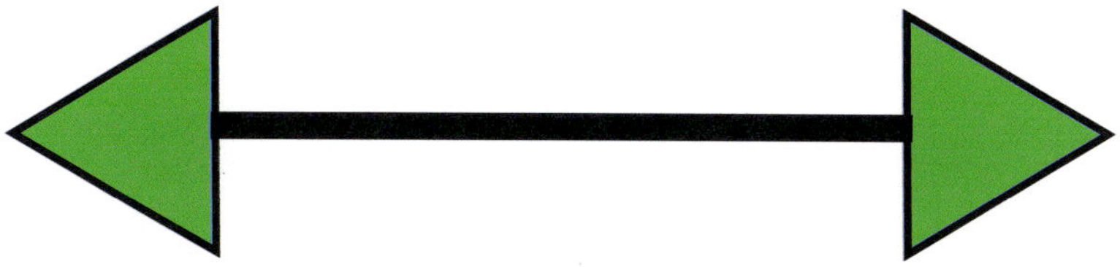

Line Segment

Ray

Ray

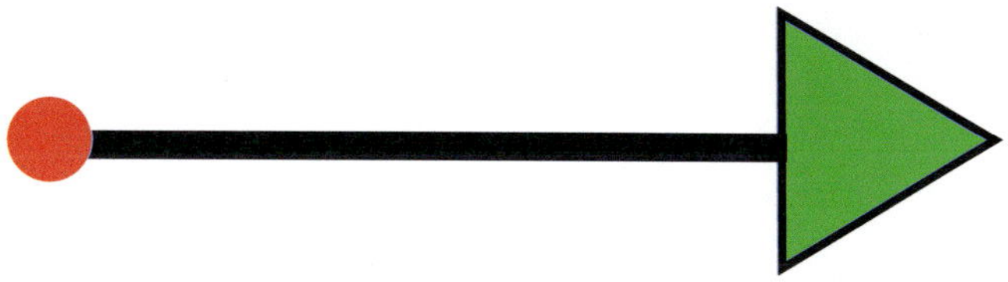

Made in the USA
Monee, IL
05 November 2026

67375722R00021